WILD ANIMAL
Book for Kids

Wild Animal Book for Kids

Coloring Fun and Awesome Facts

KATIE HENRIES-MEISNER

ILLUSTRATED BY ANDRE SIBAYAN

Z KIDS • NEW YORK

Published in the United States by Z Kids, an imprint of Zeitgeist™,
a division of Penguin Random House LLC, New York.

zeitgeistpublishing.com

ISBN: 9780593435533

Illustrations by Andre Sibayan

Book design by Aimee Fleck

Manufactured in China

3rd Printing

For Max and Luc

Meet the Wild Animals!

In this book, you will meet 25 exciting wild animals, all the way from alligators to zebras. But what makes an animal *wild*? Wild animals survive on their own without help from people; they are not tame or domesticated like our family pets.

Along with super fun facts about these animals, you'll discover each animal's classification (its category based on similar features, like reptile or mammal), its most common group name or names (like a band of gorillas), and its habitat.

So what *is* a habitat?

All animals require air, water, sunlight, food, and shelter in order to survive. They get these needs met from their environment, or habitat. Almost every place on Earth—from the hottest desert to the coldest snowy mountaintop to the deepest ocean—is a habitat for some kind of animal. Animals adapt to their habitats to survive. Some animals adapt to the climate to find food or to avoid predators. Animals like polar bears have adapted to the Arctic by developing a thick layer of blubber, or fat, to help them survive and stay warm in their icy cold habitat.

In these pages you'll color pictures set in forests, deserts, grasslands, wetlands, mountains, and oceans—all places in nature that Earth's animals call home. As you color and read, you will encounter birds, mammals, reptiles, and fish. Many of these animals thrive in a variety of habitats, so the labels you see won't always be the only places you'll find that particular animal.

Every now and then you'll see a specific type of wild animal rather than a general kind of animal (like a Burmese python instead of a snake). Sometimes there were just too many kinds of that wild animal to cover, and other times that particular animal was the most fascinating kind of animal to talk about.

You will see each animal three times: First, you'll discover (and color!) the animal along with its name, classification, group name, and habitat. Next, you'll read fun facts about that animal and color it again, this time pictured in its natural habitat. Finally, you'll find a set of bonus trading cards in the back of your book. These can be colored, cut out, and even traded with friends! Maybe you'll take yours to the zoo and add more fun facts of your own as you learn and color. Here's hoping that your coloring fun is a *wild* adventure!

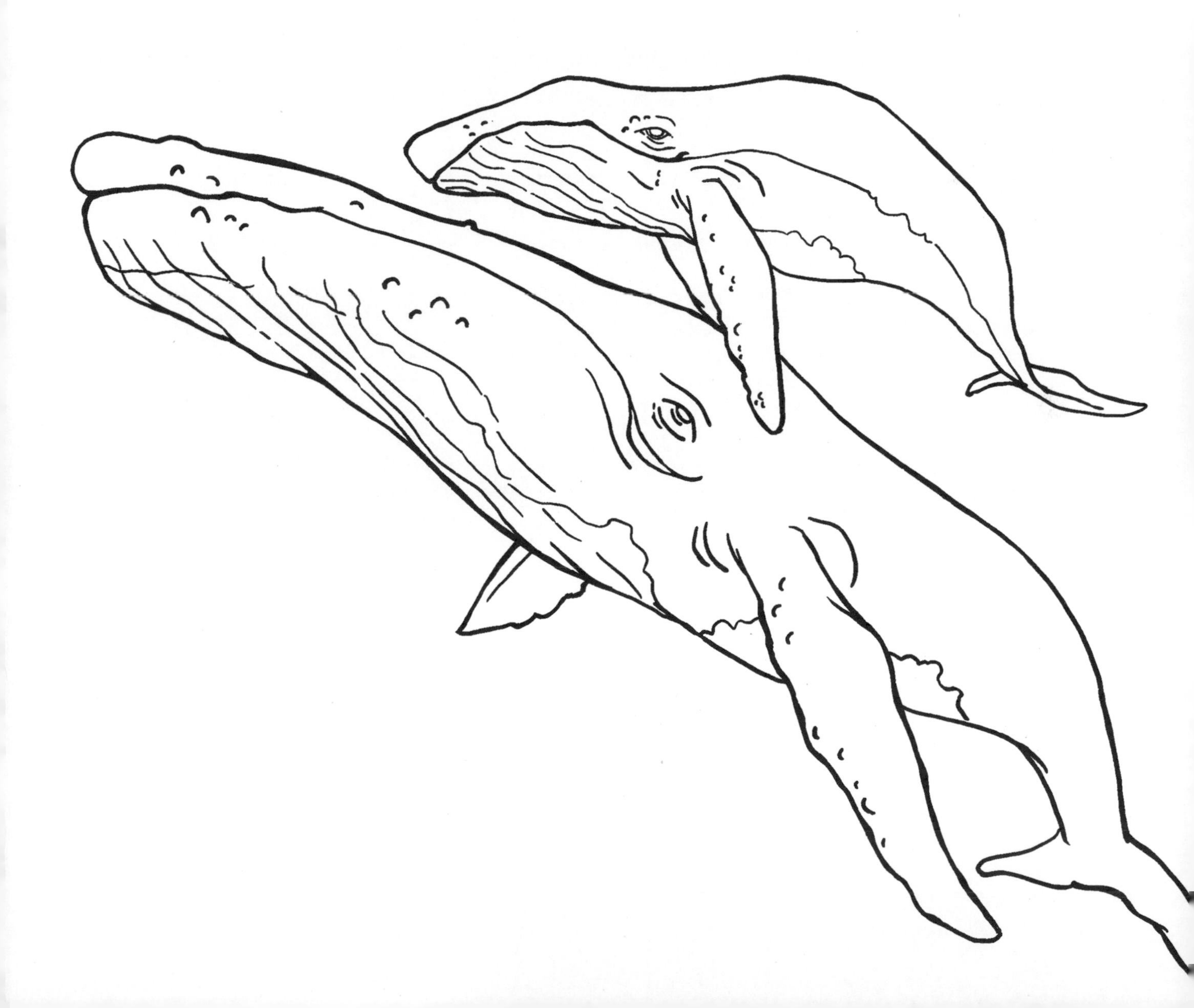

Alligator

Classification: reptile

Group name: congregation

Habitat: freshwater wetlands

Alligators are very social reptiles. They communicate through a number of different calls, including bellows, hisses, growls, and a cough-like sound called a chumpf. The American alligator is the biggest known reptile in North America. It has a long, strong tail and giant jaws loaded with up to 80 pointy teeth.

Antelope

(Gazelle)

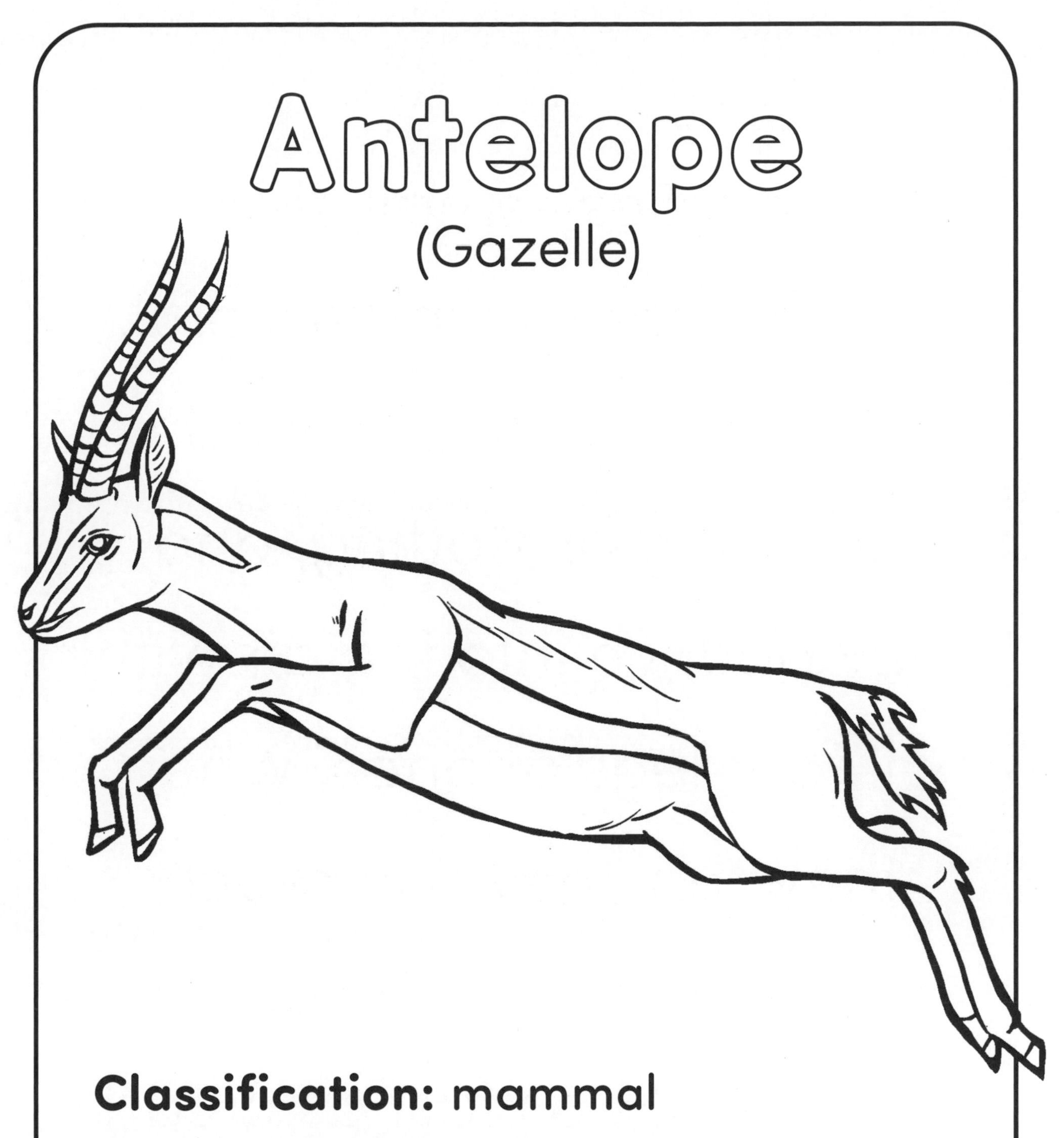

Classification: mammal

Group name: herd

Habitat: savannas, deserts, wetlands

Gazelles—small, thin **antelopes**—run in graceful, bounding leaps called pronking or stotting, and they bounce into the air with all four legs at once. When they run, they can reach speeds up to 60 miles per hour. That's as fast as a car!

(Polar bear)

Classification: marine mammal

Group name: celebration

Habitat: arctic

Polar **bears** have black skin and see-through fur. The fur appears white because it reflects light. Underneath all that thick fur is thick jet-black skin, and underneath that is a layer of blubber. Polar bears are marine, or water, mammals, just like whales, seals, and dolphins. They are great swimmers. Their scientific name means "sea bear"!

Cheetah

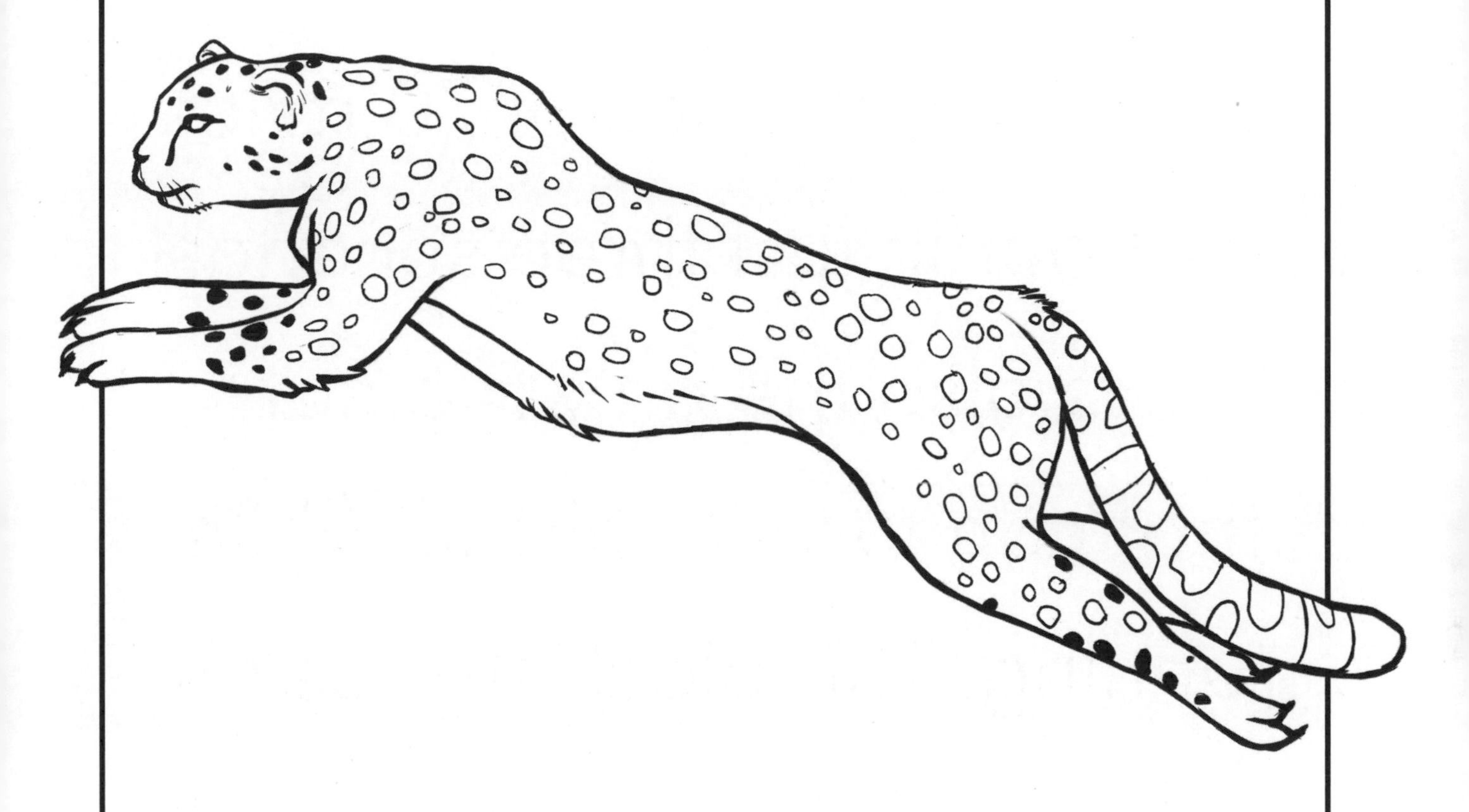

Classification: mammal

Group name: coalition

Habitat: grasslands, open savannas

The **cheetah** is the fastest land animal in the world. It can reach speeds of 70 miles per hour in just three seconds! It is also the only big cat that can turn in midair while running. The black stripes on a cheetah's face keep the sun out of its eyes during a hunt.

Chimpanzee

Classification: mammal

Group name: community

Habitat: forests

Chimpanzees live only in Africa, in a variety of forest habitats. They are very intelligent; they use more tools, in more ways, than any other animal besides humans. They also use plants to treat their injuries and illnesses. Some have even learned sign language!

Crocodile

Classification: reptile
Group name: bask (when they're on land), float (when they're in water)
Habitat: tropical wetlands, rivers, lakes, marshes, swamps

Crocodiles have the strongest bite of any animal in the world, but they don't actually chew their food. They swallow their food whole and have incredibly strong stomach acid to help them digest it. They sometimes swallow stones to help break up their biggest meals.

Dolphin

Classification: mammal

Group name: pod

Habitat: open oceans, coastal waters

Dolphins are known for being smart, but did you know that they *name* each other? Bottlenose dolphins use unique whistles to identify one another. They can also recognize their own names and the names of others. And dolphins have two stomachs: one is used to store food and the other is for digestion.

Eagle

(Bald eagle)

Classification: bird

Group name: convocation

Habitat: forests near marshes, rivers, lakes, seacoasts

Eagles use superhero eyesight to find their prey, then dive at speeds of up to 100 miles per hour to catch their meal, on both land and water. Bald eagles, the most famous eagles, build enormous nests, called aeries, in tall treetops or coastal cliffs. These nests can weigh up to 2,000 pounds!

Elephant

Classification: mammal

Group name: herd, parade

Habitat: savannas, grasslands, forests

Elephants, the largest land animals, have long, hollow trunks that can suck up water, shell peanuts, and even work like a snorkel during swims. The trunk can hold up to eight liters of water. They also have big tusks. Did you know that elephant tusks are actually giant teeth?

Classification: bird

Group name: stand, flamboyance

Habitat: lagoons, shallow lakes, mudflats

Flamingos have pink feathers because they eat lots of pink foods, like plankton and shrimp. And they eat upside down! They stir mud with their feet, scoop it up in their beak, and flip their head upside down. Then they use their beaks to get rid of the muddy water and keep the small creatures to eat.

Fox

(Red fox)

Classification: mammal

Group name: skulk, leash

Habitat: forests, grasslands, mountains, deserts

Red **foxes**, the largest true foxes, have amazing senses, including night vision and long-range hearing. They have whiskers on both their legs and faces, which helps them feel their way in tall grasses. Their forepaws have five toes each, but their hind feet have just four.

Giraffe

Classification: mammal

Group name: tower

Habitat: savannas, grasslands, open woodlands

Giraffes are the tallest land animals due to their long necks. But a giraffe's neck isn't the only extra-long thing about it: its tongue can reach over 20 inches! Scientists think giraffe tongues are black to protect them from sunburn.

Gorilla

Classification: mammal

Group name: troop, band

Habitat: forests

Gorillas are unusually caring; they show big emotions like compassion and grief. From about the age of 8 to 12, male gorillas are called blackbacks. Around age 12 or 13, a grayish-white section of hair develops over their backs and hips and they become known as silverbacks.

Hippopotamus

Classification: mammal

Group name: bloat

Habitat: wetlands, rivers, swamps

A **hippopotamus** can hold its breath for up to five minutes underwater. Its eyes, ears, and nose are close together at the top of its head, so it can see, hear, and breathe while submerged. While swimming underwater, its nostrils and ears fold shut, just like our eyelids do, to keep water out.

Lion

Classification: mammal

Group name: pride

Habitat: grasslands, savannas

Lions run fast but not for long; they get tired very quickly. They sleep up to 21 hours a day. Every lion has a unique whisker pattern on their face, like a unique human fingerprint. Male lions have big hairy manes around their head and neck.

Meerkat

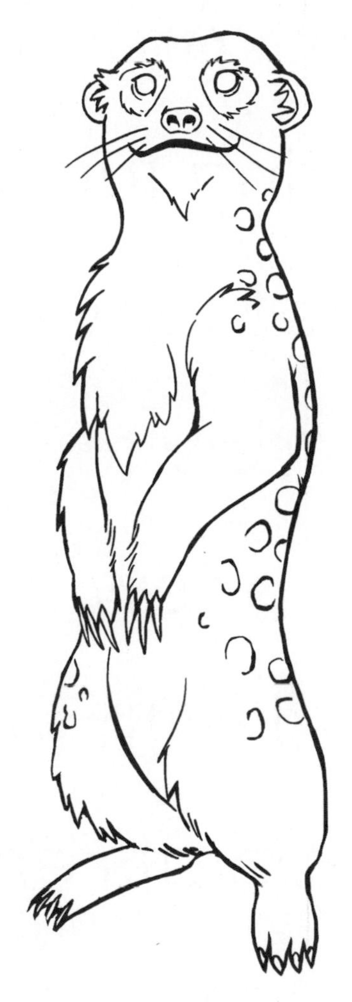

Classification: mammal

Group name: gang, mob, clan

Habitat: deserts, savannas, grasslands

Meerkats live together in burrows, underground dwellings with connecting tunnels, rooms, and exits. They leave the burrow during the day to forage for food, while a few babysitters stay behind with the young. When meerkats encounter a predator, they stand close together and hiss loudly so they look like one large animal.

Classification: bird

Group name: parliament

Habitat: forests, grasslands, mountains, deserts

Owls have superpower hearing. They can even hear animals under the ground—up to 12 inches beneath the snow! Some owls have ears at different heights so they can hear better from different directions. Their small, flat faces have special feathers that help direct sound waves into their ears.

Penguin

Classification: bird

Group name: colony, rookery

Habitat: oceans, coasts

Penguins are known for their black-and-white tuxedo, which is actually a kind of camouflage. When swimming belly-down in the ocean, their dark backs blend in with the color of the water so predators can't see them from above. Their light bellies make it hard for predators to see them from below, too.

Classification: mammal

Group name: crash

Habitat: grasslands, wetlands, forests

Rhinoceros horns can grow to be over four feet long. But they aren't made of bone—they're made of keratin, like your hair and fingernails. If a rhino's horn breaks off, the rhino can grow a new one!

Seahorse

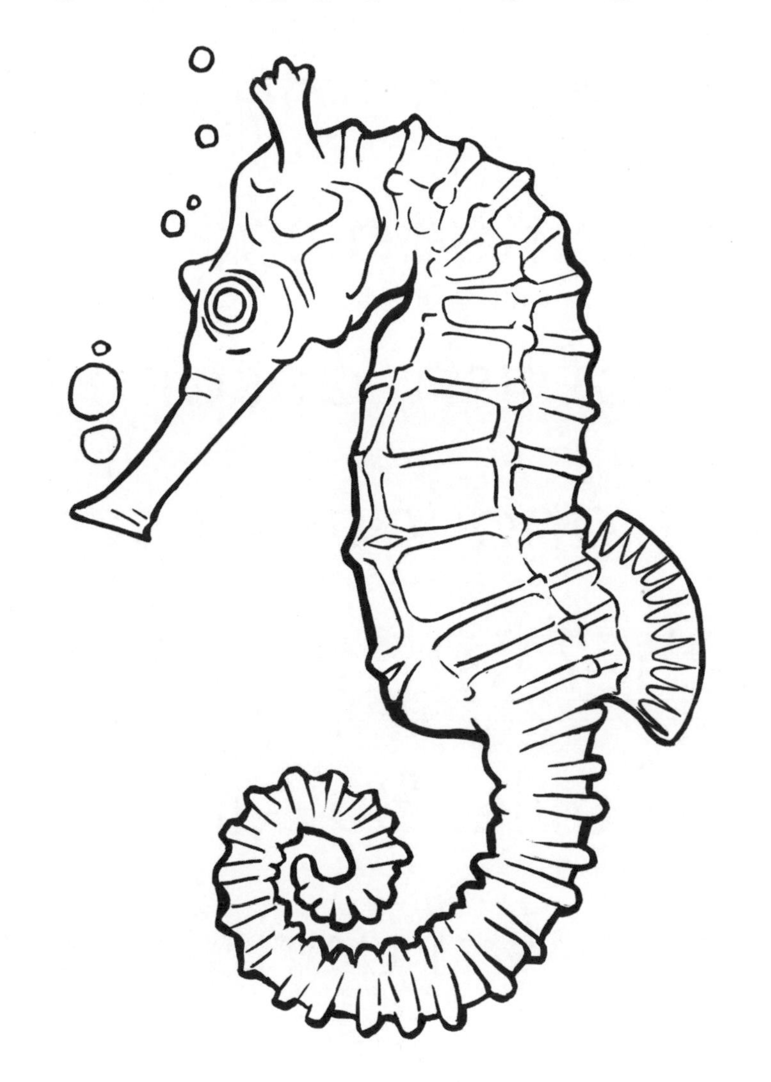

Classification: fish

Group name: herd

Habitat: oceans

Seahorses—named for their horselike head shape—are a kind of fish, despite their shape and lack of scales. Like other fish, they breathe through gills. But unlike most fish, they swim upright, among water plants and seaweed.

(Hammerhead shark)

Classification: fish

Group name: shiver

Habitat: oceans

The hammerhead **shark**, one of the most unusual sharks in the ocean, has a wide, flat head with eyes on either side. This means it can see above and below . . . but not right between its eyes! Scientists believe its head shape helps it locate and capture stingrays, its favorite prey, on the sandy seafloor.

Snake

(Burmese python)

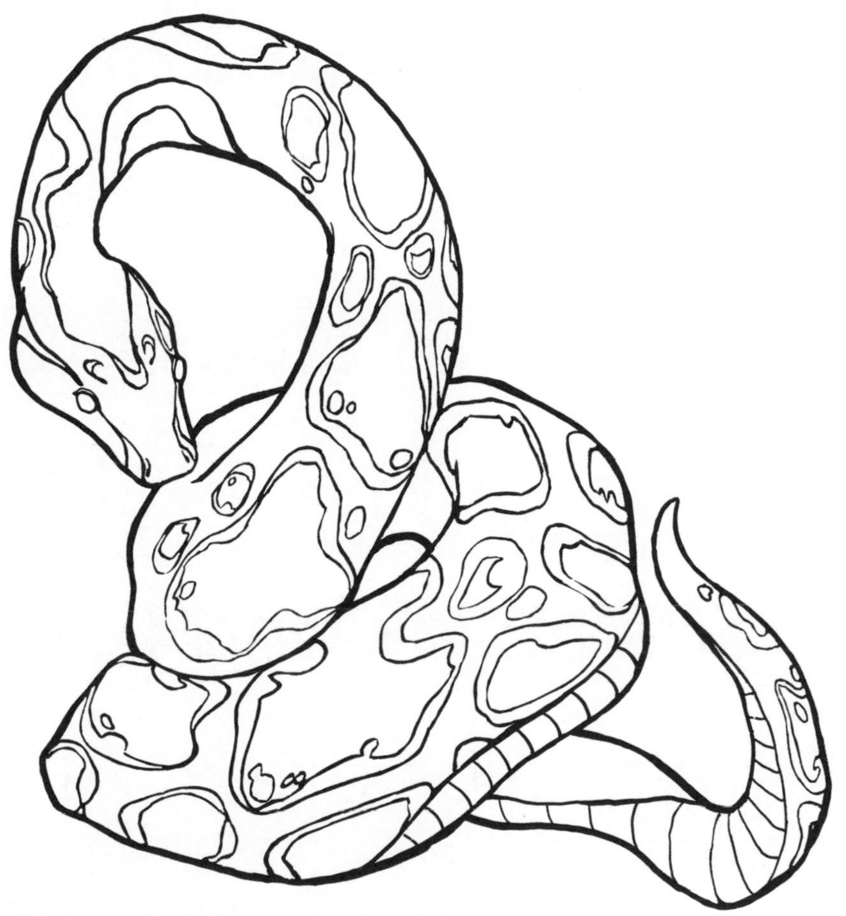

Classification: reptile

Group name: pit, nest, den

Habitat: swamps, wetlands, grasslands, marshes

Burmese pythons are among the largest **snakes** on Earth. They spend a lot of their time in trees. Like other snakes, they don't have eyelids, so they don't blink and they sleep with their eyes open. And even though they have nostrils, they smell with their tongues!

Tiger

Classification: mammal

Group name: ambush, streak

Habitat: forests, grasslands, savannas

Tigers are the only cat species that is striped all over; even their skin is striped! No two tigers have the same stripe patterns. They have excellent night vision and hunt alone, largely in the dark. They are also great swimmers and can swim long distances to hunt.

(Blue whale)

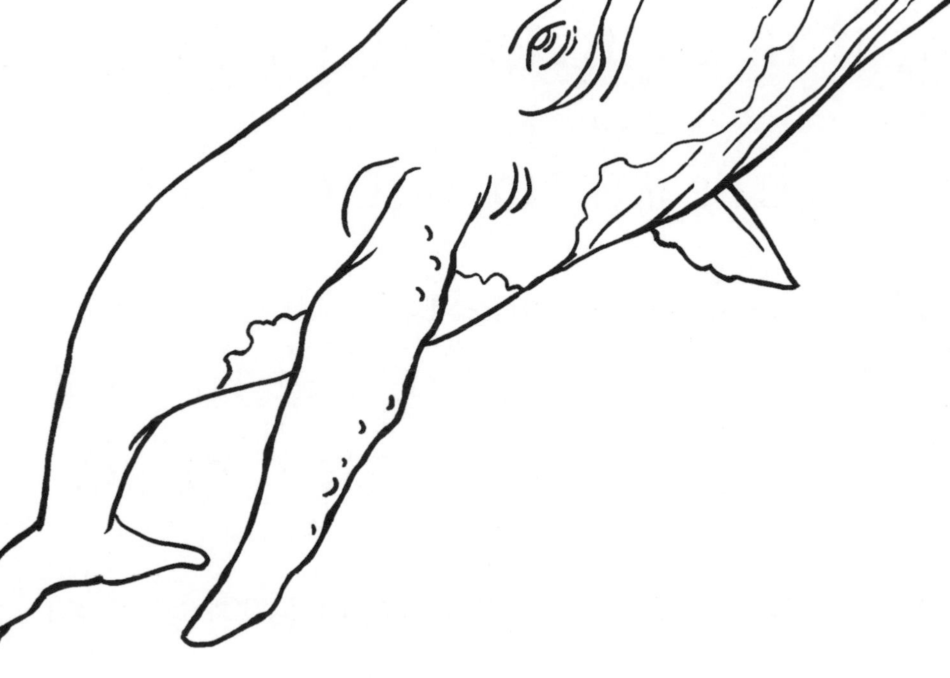

Classification: marine mammal

Group name: pod, school, gam

Habitat: oceans

The blue **whale** is not only the largest whale, it is the largest animal *ever* on Earth; it is bigger than even a dinosaur! Its tongue alone weighs as much as an elephant. Blue whales also are the loudest creatures on the planet. Their calls can travel 1,000 miles or more.

Classification: mammal

Group name: dazzle, zeal

Habitat: grasslands, savannas

Zebras are known for their stripes, which are a form of pest control almost like built-in bug spray. Scientists believe the black-and-white striped pattern helps them avoid being bitten by flies. A zebra's stripes are also unique to that zebra; no two patterns are the same.

About the Author

Katie Henries-Meisner, author of *Dinosaur Book for Kids: Coloring Fun and Awesome Facts* and *Bug Book for Kids: Coloring Fun and Awesome Facts,* is an educator-mom of two wild kids: Thalia and Harrison. She's from Massachusetts, where she grew up with three sisters and started her career as a teacher. Since developing a passion for education and social justice, along with a love of learning through exploration, nature, and projects, Katie continues to work in education and now lives in Napa, California. She is a lifelong lover of the written and spoken word, and sometimes talks too much.

About the Artist

Andre Sibayan is a multi-disciplinary creative from the San Francisco Bay Area. He also illustrated *Dinosaur Book for Kids: Coloring Fun and Awesome Facts* and *Bug Book for Kids: Coloring Fun and Awesome Facts*. He dedicates this book to his late mother, Virginia (Beth) Sibayan, who stoked his curiosity for the natural world with hours upon hours of nature shows.

Alligator

Antelope

Bear

Cheetah

Chimpanzee

Crocodile

Dolphin

Eagle

Elephant

Flamingo

Fox

Giraffe

Gorilla

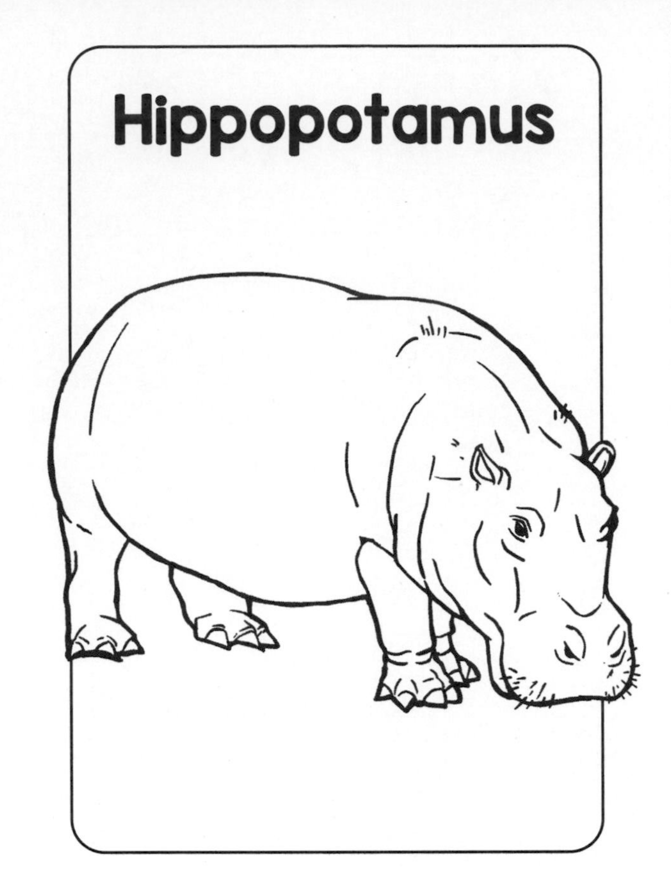
Hippopotamus

Lion

Meerkat

Owl

Penguin

Rhinoceros

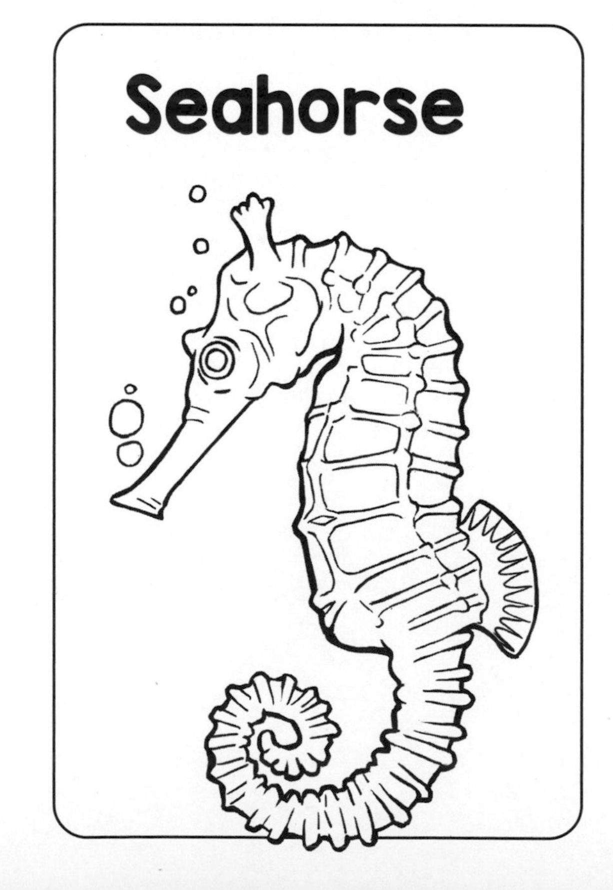
Seahorse

Shark

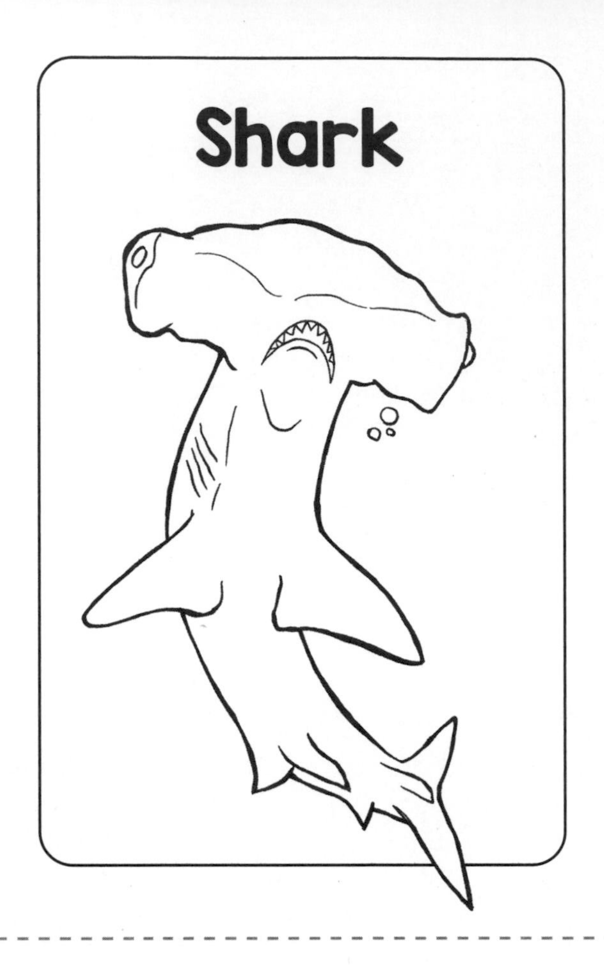

Snake

Tiger

Whale

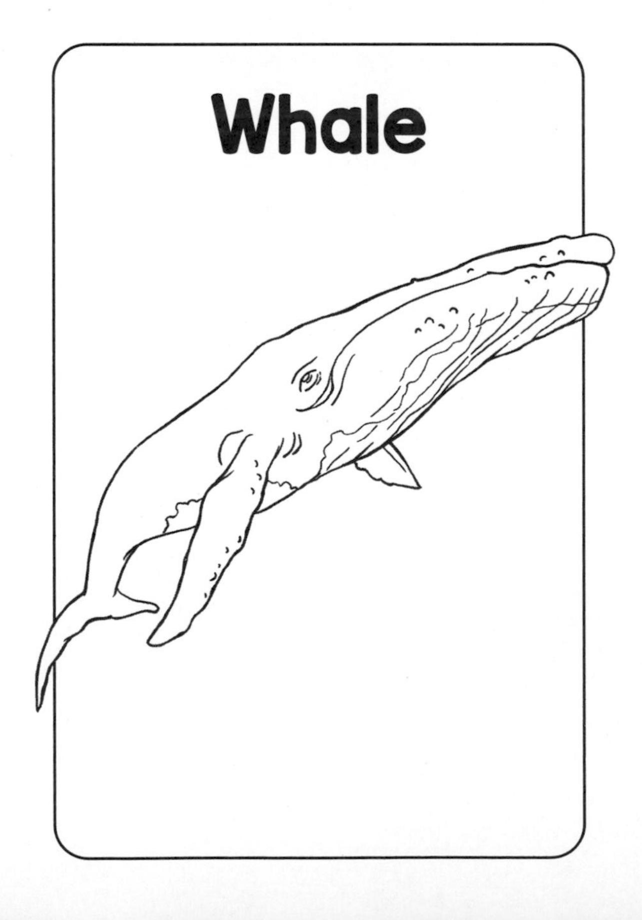

Zebra